Antoni Lacinai is an international expert on workplace communication, motivation and engagement. His business passion is to decipher the mysteries of human communication, especially around leadership and customer communication.

Antoni has coached thousands of people on how to meet, greet, and treat customers when working at trade shows. When large international organizations or national governments need to elevate, for instance, their leadership, sales, and presentation skills, they reach out to Antoni.

To date, he has written and co-written 14 books. He is an international keynote speaker and is often interviewed on Swedish national TV.

I dedicate this book to all my wonderful customers. It's an honour and a privilege to contribute to your missions and goals.

Antoni Lacinai

SPEED SELLING

How to Generate Leads at a Trade Show

AUSTIN MACAULEY PUBLISHERS™

LONDON • CAMBRIDGE • NEW YORK • SHARJAH

A CIP catalogue record for this title is available from the British Library.

ISBN 9781035823086 (Paperback)
ISBN 9781035823109 (ePub e-book)
ISBN 9781035823093 (Audiobook)

www.austinmacauley.com

First Published 2024
Austin Macauley Publishers Ltd®
1 Canada Square
Canary Wharf
London
E14 5AA

A big thanks to Åsa Dahlqvist, who was my writing partner for the first Swedish version of this book. Åsa has changed careers since then, but without our collaboration, this book would not have seen the light of day.

Another big thanks to Austin Macauley Publishers, who saw the potential in me and the script and made the book much better than what I could have done myself.

Table of Contents

Foreword
by Antoni Lacinai

Foreword
by Roger Kellerman

I have dedicated my work life to the world of business events, fairs and exhibitions, congresses, conventions, conferences and incentive travel. I have lived and breathed this beautiful industry for the last 40 years. The meeting industry is huge. Globally it employs 27,5 million people and it is a 1,6-trillion USD business and will have an estimated growth rate of 12 % every year between now and 2032, according to Custom Market Insights (CMI). Last year, 6 billion people participated in business events across more than 180 countries, according to PCMA.

Those are the hard facts, and it means that it should be taken seriously.

Not only for the hard facts though, but also for the sheer human aspects. You see, we humans have to meet. It is part of our DNA. We cannot exist in isolation. We need other people to grow, spiritually, mentally, physically. This is how we share ideas, thoughts, innovations. This is how we sell, serve, and collaborate.

Trade shows and exhibitions are often a part of conventions and conferences. Or they exist in their own right. Imagine in ancient times where people displayed their products at different markets. They sold fruit, crops, and live stocks. We still do this today, but it can just as well be t-shirts, machines, or it-systems. Trade is happening all over the world and many businesses use the opportunity to show what they can offer to customers and prospects at different trade shows.

I got to know Antoni back in 2006 when he was writing his first book, which was about presentation skills. Communication has always been a passion for Antoni, and this book, Speed Selling, is a gem for anyone who want to perform well and get more business from exhibiting. Too many people in the booths I visit, or pass by, seem uninterested and bored. This is not helpful for them, for the visitors or for the trade show at large. We can do better.

I believe this practical and very useful book should be given to each and every one who will represent their company or organization in a booth. The fact that you are reading this gives you a competitive edge to the other ones who don't. Especially if you practice what you will soon learn.

I wish you a great read and a great upcoming exhibition.

Roger Kellerman
Publisher
Meetings International Magazine
Winner of JMIC Profile & Power Award 2014

Exhibitions, Trade Shows, Fairs...

There are thousands of meetings happening all around the world right now. A lot of these meetings are different kinds of trade shows, with exhibitors who have everything from a roll-up and a high desk to booths that cover several thousand square feet.

Trade shows are fascinating. One day you have 50,000 telecom nerds gather and talk about apps, security and the latest products. Afterwards, everything is torn down in a few hours, to then house 25,000 flower lovers who want to know more about plant protection, nutrition, stamens and pistils. Then everything is taken down again, to showcase the next topic to a crowd of enthusiasts in another trade. It is all done to meet others, cultivate relationships and make business.

Why should you read this book?

As booth staff, you need the right preparations, training and education to maximise the possibilities that the exhibition offers.

There is a difference between standing in the booth and working in the booth.

We want every meeting at the exhibition to meet the expectations that the potential customers has, and we want every meeting to generate new energy, knowledge and, eventually, more business. We also want you, working in the booth, to feel safe in your role at the exhibition and make sure that you are prepared to create lots of optimised meetings within the limited opening hours of the exhibition. That is what speed-selling is all about.

Who are you?

Whether you are a manager, a sales rep, a product expert, a marketer, a communicator, an administrator etc., your participation in the booth is important to get a positive return on the investment of time and money.

This book can also be used as a handbook for you, the project manager—responsible for the setup, the daily running, and the performance of your co-workers hour by hour, day by day. There is a checklist added at the back of the book as an appendix of how you, the one in charge of the exhibition, can prepare your staff.

If you are the exhibition organiser, you will also stand to benefit from reading this book. To keep the exhibitors satisfied, their staff needs to be well prepared and able to create those valuable person-to-person encounters that make the show worthwhile...and worth the investment in a booth next year too.

This book is for you who want to get more leads at the trade show.

In this book, we will deal with how you as booth staff can act and how you can build business opportunities during the exhibition. We believe that this *booth behaviour* is the most important factor to get good results at the exhibition. It doesn't matter how well designed your booth is if the people staffing it don't know what to do.

But naturally, it will help a well-prepared staff if you, the overall project manager, have paved the way by creating a well-functioning booth, inviting visitors and adding activities that attract visitors. The work of the booth staff will, moreover, be in vain if there is not a plan of how to get back to the new contacts after the exhibition is over.

There is a checklist—"How to plan an exhibition participation" in the back of the book.

Who are we?

We, the writers of this book, train people and give lectures in communication, motivation, goal-setting, meetings, exhibitions, networking and mingling. We know just how much better an exhibition could be when you and your colleagues are given tips and tools on how to be sharper in the booth. We know this because we have prepared and coached several thousands of people and we have seen the positive effects of gaining new knowledge and training.

Antoni Lacinai is a global speaker and a communication & motivation expert working at the cross-point of leadership, employee engagement and customer experience. One of his favourite areas is to coach exhibition staff. To date, he has written 14 books, tons of columns, given a TEDx talk and is frequently on Swedish National TV.

You can find him on LinkedIn, Instagram and at antonilacinai.com.

Åsa Dahlqvist was one of Sweden's leading experts in exhibitions and a highly appreciated lecturer and trainer in planning of exhibitions, booth behaviour and networking. She loved to help exhibitors have more and better meetings. Note: Åsa is no longer in the business. She went to an employment and has left the consultant world.

Remember: practice makes perfect! Be sure to use the tips, again and again, till they become a habit. Keep referring to the checklists. Rehearse. Seek feedback and give it—constructively—to your booth buddies, both before and after the exhibition. Have fun! That's the route to exhibition excellence.

Working as Booth Staff

You are the most important asset at the exhibition!

There are lots of different reasons to exhibit:

- Direct sales
- To establish contact with new customers
- To strengthen relationships with current customers
- To create interest for the offer
- To increase the understanding and the knowledge of you and your offer
- To strengthen your position on the market
- To receive feedback on a product or a service
- Recruitment

Whatever your priorities, the encounters between the exhibitor and the visitor are the heart of the matter. The booth staff have the responsibility to create and conduct those meetings—to welcome, persuade, explain, listen, educate, demonstrate, show interest and display what your organisation stands for and who you are.

A good-looking booth can't do the job for you. Nor can fun activities or nice giveaways.

It is you and your colleagues who, together, create the value of the exhibition and it is you who are the most important ambassadors for your brand.

There is often a mixture of salespersons, various experts and assorted support staff in the booth. It's like a national team where the best players of every local team are called upon to make a temporary effort together—and every individual has their own vital task.

In a typical booth, we might find:

- Product experts
- Product managers
- Executives (visiting, overseeing, evaluating…)
- Technical support
- Marketing people
- Salespeople
- Assistants
- Hired booth hosts

Even if the booth staff have great knowledge of the product, there is often a lack of skill—how to behave in the booth and what it takes to do a really good job at the exhibition. It is not easy to reach out to strangers and engage effectively in an interesting conversation, while promoting your product. As is so often the case, preparation and training are required.

Every person in the booth represents the brand

A while ago I visited a booth that made a very good impression on me. The theme was football teams. They had a competition where you could try and score with a football, and their message was tied to words like "team", "teamwork" and "supporters". The booth staff looked happy and nice and seemed to be active in the booth. I received a ticket to the VIP-section of the booth, where they handed out hot dogs and sodas. But when I tried to order my hot dog and curiously asked what the green ketchup was made of, all I got was a sigh from the girl who was serving the food. She rolled her eyes and clearly showed how tired she was of getting that question, and then pointed towards a little note on the counter while curtly saying: "You can read about it there."

I left the booth a little disappointed, feeling I had been poorly treated. Even though I did understand that there was only one person who had been rude, my entire view of their company and brand had been damaged.

Everyone working in the booth affects the brand and how your visitors perceive you, whether your job is to hand out candy or make a sales pitch.

Åsa

Are you the master of exhibitions?

When we ask booth staff before a trade show what characterises a good booth staff, we receive the following answers:

- Friendly
- Active
- Welcoming while greeting visitors
- Not too pushy or forward
- Treat all visitors well
- Filled with energy
- Listening
- Empathetic
- Curious and interested
- Knowledgeable
- Have arguments that meet my needs
- Articulate
- Trustful
- Focused
- Explains in an easy way
- Committed
- Helpful

The next time you visit an exhibition, see how many of the exhibitors show these qualities and behaviours. You can use the Behaviour Bingo in the appendix to this book. Which good quality is the most common? Which behaviour seems to be the hardest to display? What can you do to "fill your Bingo card" the next time you work at an exhibition?

Don't be like Snow White and her seven friends...

Based on our observations and our experience, there are very few exhibitors who truly act in the best way possible. Here are some common mistakes, illustrated through Snow

White and her seven little friends. Do you recognise them? And have you not yourself, from time to time, acted like some of these characters?

Snow White has put on her most beautiful dress and is now standing, passively waiting for the prince to contact her, instead of going to look for him. Unfortunately, this often leads to many missed meetings and potential contacts—not an effective way to work in the booth. And to Snow White, it also feels very uncomfortable and awkward, just standing there and trying to look nice—it can even seem so artificial that visitors choose not to stop at her booth and simply pass her by.

Doc already knows everything. Listening to the visitor is redundant, since he has advance knowledge of what is required… or at least of what he is selling—and soon the visitor will have the information too. No value in this, no engagement. It will be a fruitless, one-way communication.

Bashful. We don't see a lot of him in the booth. He prefers to hide behind the panel, a table, a curtain…or other people. He doesn't like to leave his comfort zone, even though he knows he should, and he always makes sure that he has something else to do in the booth rather than greeting visitors. He checks his phone, tidies up the pamphlets, refills the candy bowls, wipes off the tables. But when you busy yourself with petty matters, the visitors choose to walk past…

Happy is a socially competent and agreeable person, who makes everyone feel comfortable. But when the visitors leave, they can't really remember what they gained from the visit in the booth, other than a few pleasant minutes of chat, some bits of candy and a pen with Happy's company logo.

Grumpy is leaning on a table, or even sitting on the sofa, and when only an hour has passed he glances at the clock and sighs. "Isn't it time for lunch soon?" Grumpy has neither the energy nor the motivation to work in the booth. He doesn't like trade shows, he doesn't understand their purpose, and he is sick of all the inquisitive visitors. And his feet are starting to ache... No visitor feels like entering a booth to cheer up a Grumpy.

Sneezy is a true hero. Even if he is sick, he will drag himself to work. He sniffs and coughs and catches his sneeze in his hand before extending it to greet the visitor.

Sleepy likes exhibitions. And even more so at night, after the exhibition has closed for the day—party time until morning comes! With a stifled yawn and an unpleasant morning breath, he greets the visitors while longing for a cup of java—and the next party.

Dopey thinks that the most important part of the exhibition is to have fun with his colleagues, or to play pranks on the staff in the neighbouring booth. A crowd of booth workers is gathered around him, joking and laughing. But the visitors don't stop by the booth. They don't want to interrupt Dopey and his friends who seem to have such a good time hanging out together.

How come we make mistakes?

Why do we keep seeing Snow White and her seven friends in booth after booth, at exhibition after exhibition, when everyone knows how to behave as staff in a booth? We have concluded that it basically comes down to the following three factors:

- The booth staff don't feel motivated enough; their role—their job in the booth—is unclear, and that makes them feel insecure within the team.
- The booth staff simply do not have what it takes to persevere—physically and mentally.
- The booth staff aren't prepared to interact with the visitors—they simply haven't received the education and training they need to create many good meetings.

Your exhibition investment is mostly in vain if you and your colleagues are not properly prepared. And it is truly an investment to participate at an exhibition: the space rent, designing and building the booth, demonstrations, product samples, pamphlets, staff, travel, logistics and food... If all the invested money and resources are not used properly, not only do you waste the investment, but there is also a risk that your brand will be damaged instead of being strengthened.

Are you prepared for your assignment at the exhibition?

Here are a few questions for you. If you answer "yes" to all questions, then you are well prepared. If you answer with a lot of "maybe" or "no", then there is still a lot of work to be done—which is the case for most exhibitors. (The test is also added as an appendix in the back of the book.)

- Do you know why you are taking part in this exhibition? What is the purpose?
- Do you know what goals you have during this exhibition?

- Do you know what your task is in the booth?
- Can you explain what is unique about your products and solutions?
- If there are many of you in the booth: Have you got to know one another? Do you know your colleagues' names? Do you know what special skills they have and when it is suitable to guide the visitor to them?
- Do you know how to make a good first impression?
- Do you know what to do to keep yourself filled with energy throughout the entire trade show?
- Do you know exactly which visitors you want to spend some more time with? And who you should just greet politely, show the goods, and say Goodbye? Do you know how to find out who you have in front of you?
- Do you know how to make first contact with a visitor without being too pushy?
- Do you know how to get the information you need to be able to give the visitor a solution?
- Do you know when it is time to end the conversation and seek commitment from the visitor?
- Do you know what to do after the exhibition to keep your promises? Have you set aside time for the follow-up?

Doing it right at the exhibition!

Crucial considerations:

- If you know the purpose of your participation—why you are there—your motivation will be greater…more energy and perseverance!
- If you know what goals you have set, it will be easier to know what needs to be done. You and your colleagues will move in the same direction, give each other feedback (praise and constructive criticism), and celebrate each successful step along the way.
- If you know the difference between a successful meeting with a customer and a time-filling talk that leads nowhere, this awareness will transmit itself to the visitor. Even if there is no direct sell or handshake, the seeds of a good relationship can be planted.

When booth staff are well prepared, you will have happier visitors and a happier staff. You will have more visitors and more meetings and reach a better result. Many visits make your booth a magnet which attracts more visitors. It will be the place where people want to be. Your brand will be strengthened. You will do everything to maximise this unique opportunity to meet many established customers and also potential clients, all at once. The relationships you make at the exhibition will live on, long after the exhibition days are over. If used properly, it will become a huge advantage.

People attract people.

In the rest of the book, we will give you lots of tips and advice on how to better prepare yourself and your colleagues to become the masters of exhibitions.

The first part deals mainly in what you need to do before the trade show. We begin with a chapter about purpose and goals—that is the foundation of all the work in the booth. Then you will receive tips and advice on how to practically prepare to feel safe and have enough energy to work in the booth.

Following that, in the second part, we will address how to best act in the booth, to maximise the number and quality of meetings. Finally, we will talk a bit about what you ought to do after the exhibition, to reap the rewards of the work you set in motion at the exhibition.

Before

Purpose and Goal

Why are you exhibiting and what is expected of you?

Clear goals give you direction and motivation in your work at the exhibition. It's as if you and your colleagues are going on a rowing trip together. You have to:

- understand why you are getting into the boat
- know where you are going
- know what role each person has on board, and how that contributes to the successful journey
- know how to use the oars properly
- know what strategies you can use, depending on weather conditions, to finally reach your destination

In other words: you need to know WHY you have a booth and WHAT you need to accomplish before you can decide on HOW to get there.

If you know the Why and What, the How will come easily!

When you know why you are there and what you are expected to achieve, it will be much easier to understand your task in the booth. You will not stand idly by, waiting for something to happen, like so many people in other booths.

The purpose: Why are you at the exhibition?

If you have not found out why you are at the exhibition, ask the project manager. But don't settle for answers like: "We are usually here", "The CEO thinks it's a good idea", "Everyone else is here...", instead, ask the Why-question a few more times. Search for the progressive and positive wording which tells you where you want to go, who you want to be in your field of work, what you want to eventually accomplish, for example: "We are at the exhibition to show people that we are one of the leading companies in the field", "We are here to strengthen the relationships with our clients" or "We are here to find new, potential clients."

Understanding the purpose gives energy,
motivation and persistence.

The involuntary exhibitor

I once asked an exhibitor, who I met at an exhibition a couple of years ago, what their intention was in being there. "Because we have to," he answered. I observed his colleagues in their well-designed, but empty, booth. They were standing there with bored faces, waiting. I asked if he was pleased with the exhibition and he just shook his head. No, obviously they didn't have a great exhibit, but what could they do, they had to be there, he complained.

Let us reflect about this phenomenon. How can an exhibitor "have to" exhibit? Needless to say, no one was forcing them. The 'have to' was a fear of 'if we don't...'—typically, 'If we don't show up here, the industry will think we're in trouble.'

What if they had chosen to turn the negative purpose into a positive purpose instead—the other side of the same coin: "We exhibit to strengthen the relationships with our current clients and to show them that we still are one of the top three in our field." Then the booth staff would feel motivated to reach out to the visitors instead of feeling that they have been handed a meaningless task. They would be proud of being a part of the company's position and the building of their brand. That is how important it is to have a progressive and positive purpose to your participation in the exhibition. Otherwise, you might as well just stay at home.

Åsa

The goals: What do you aim to achieve?

The psychology of goal-setting is well understood. If you set the right kind of goals, your efforts are boosted by some 15-30%, according to the goal experts and writers Edwin Locke and Gary Latham. Those are big numbers, so be sure to find out what you are expected to accomplish at the exhibition. What will you have accomplished when the exhibition is over?

Both purpose and goals need to be set early in the planning process—even before you, as booth staff, have been involved in the project. But please, remind the project manager that you need to know what is expected of you and how it will be evaluated. '

If you are involved in setting the goals, it may be good to know what sort of goals work well at an exhibition.

The overall goals should be linked to your exhibition purpose and can be specified through questions like:

- How many leads do we want during the exhibition?
- How many visitors do we want in our booth?
- How many products do we want to sell?
- How many of our VIPs should visit our booth?
- How many should give us feedback on our new product?
- How many should enter our competition?

The criteria for a good exhibition goal are the following:

- The goals should be valuable to you
- The goals should be tangible and measurable
- The goals should be relevant

Find out what goals you have for your participation at the exhibition and make sure to find out what is expected of you to fulfil these goals.

How do we reach the goals?

Before you take your place in the booth, you also ought to know what your daily goals are. That is something you need to agree on together. It is just like a football team; not all players score but everyone's efforts contribute to the goal.

And you should also be ready to improvise, switching positions and roles with each other—in the booth, just like on the field.

Now we get to the next interesting question: HOW do we reach the goals together? The strategies depend on the exhibition, the exhibitor, the target group. But there is one common denominator among all exhibitors: they never reach their goals unless they create as many meetings—good meetings—as they possibly can during the exhibition.

It's up to each individual in the booth to engage with every visitor. If the visitor is not interesting, then end the conversation quickly. If the visitor is a customer, a potential customer or an interested party, you must have strategies to find out about their needs and challenges, and then to get them interested in further contact with you, or even buying your products and services. And this is what the rest of the book will cover.

Prepare to Perform

Feeling secure

Feeling secure in the team and having the right conditions to keep your strength up are important, so you can perform at your best during the exhibition.

To be able to feel secure and safe in the booth, do your research in advance. The checklist below can also be found as an appendix in the back of the book. Some things will apply to you and some will not, so pick out what is relevant to you—for the next exhibition—and set the rest aside.

- Do you have new products that you are introducing? What are they? What do you need to know to be able to present them well?
- Do you have a new slogan, a new brand, a new service…or something else whose background you should know?
- Have you invited any VIPs? Who has RSVP'd?
- What is your policy about journalists coming to your booth to interview you?
- What policy do you have concerning competitors that visit your booth?

- What policy do you have for people visiting your booth, who are not in your target group?
- Will any of your top executives be there?
- Is there a plan on how to treat press, analysts and other groups of visitors that are not regular clients?
- Are there other things going on around you that you ought to know about? Lectures, conferences, competitions, and so on?
- What does the trade show look like? Where are other key companies in the field located? Where are the restaurants, cafes and toilets?

Feeling safe in the team

Lars Lagerbäck, the former manager of Sweden's and many other countries' national football teams, was interviewed in a radio channel, about how important it was for the players to get to know each other. His words: "Those who have good relationships off the field, play better together on the field."

The same goes for a booth staff at an exhibition. The better you know each other, the safer you will be, and you will more easily share questions and tasks with each other. And you will be better at tackling setbacks and finding solutions to problems.

Here are a few things you can do to build a better team spirit and a sense of community:

- Get to know each other. Unfortunately, there are many who try to do this during the exhibition, so we find the booth staff standing around talking to each other instead of talking to the visitors. The optimal setup is to have a joint meeting before the exhibition opens, when you can review the itinerary of the day, but also have an opportunity to mingle and get to know one another. Everybody should be encouraged to grab a coffee or lunch with those they don't know so well.

- Help each other and leave the rivalry at home. You are a tight team with a common goal during a limited period. There is no time for bickering among yourselves.

- Do "buddy-checks" on each other: Is my shirt untucked? Is my fly open? Does your colleague have spinach in his teeth? The word "buddy-check" comes from diving, where the rule is to never dive alone. Your buddy checks your equipment so that everything works. Then you check your buddy's equipment.

- Celebrate when you have reached a goal, when someone does something extraordinarily well or if it is the birthday of someone in the team.

How can you contribute to the booth staff having a good team spirit?

Theme song

Why not have a cheerful song that you play every time you have a meeting before the exhibition? One of my clients did that. They chose a song and played it at every meeting they had the weeks before the trade show. It became a positive trigger. When the trade show opened that first morning, they did the same thing. Everyone felt ready and boosted to begin.

Antoni

Practical Tips Before and During an Exhibition

To be at an exhibition is demanding, both physically and mentally. If your body and brain don't have enough energy, you really can't perform well. Here are 12 practical tips to help you.

Practice standing. If you have a job at a desk, your feet are in for a shock if you don't start standing at work, in the run-up to the show. Start a few weeks in advance. Stand and walk around more than you usually do.

Take a break. Make sure to take a break every couple of hours. You need to rest both your legs and your head. The optimal setup would be to have an agreed schedule you can follow. If you don't have one, it is, of course, important that you check in with your colleagues before leaving the booth.

Recharge your batteries during your breaks. Do not wander around the trade show, visiting other booths. Leave the exhibition hall and recharge with what you need at the time: food, drink, a comfortable chair, fresh air…

Wear comfortable shoes. After a few hours a pair of shoes can pinch. A new pair to change into could be a good idea. Make sure to break in new shoes before the exhibition.

Change your socks. We sweat approximately 4 cl a day per foot. If you change socks, it will feel better. There are also compression socks that help with the blood circulation.

Breathe—the right way. In pressured situations, our breathing accelerates and we do not fill our lungs to their maximum capacity. That makes us even more stressed out. Breathe slower to feel better.

Take good care of your voice during the exhibition. Try to lower your pitch, and talk with the support of your stomach, instead of relying on the upper part of your vocal chords. That way you will not strain your system so badly and you will have a deeper and more confidence-inspiring voice.

Drink water. Even if the room is cool and you don't sweat a lot, the air is usually very dry in the big, air-conditioned halls. Get in the habit of taking a sip of water between conversations. Your breath will also stay fresher and it will help your vocal chords.

Think of what you eat. Remember that a shift in a booth is as tough as any other physical workout. You need to stock up on energy. Remember to eat a proper breakfast—things that are filling, but not indigestible. Eat fruit and snacks occasionally, but avoid sweets that get your blood-sugar levels going.

Hold back on the alcohol the night before. Do not party hard at night. Think of it as competing in sports the day after. How would you prepare yourself then?

Have fun. Be in a good mood. Give each other a high five, hand out compliments, laugh and feel good.

Do a power pose every morning. According to a TED talk with the social psychologist Amy Cuddy[*], you can, by standing like an invincible superhero for 120 seconds, lower the level of your stress hormone cortisol and raise the level of the power hormone testosterone. You will become stronger, more sure of yourself and better focused.

[*] Observe that Amy Cuddy's research has been hard to replicate by other scientists. Perhaps it's only a placebo effect. Sometimes though, that's all you need…

During

The Meeting Wheel

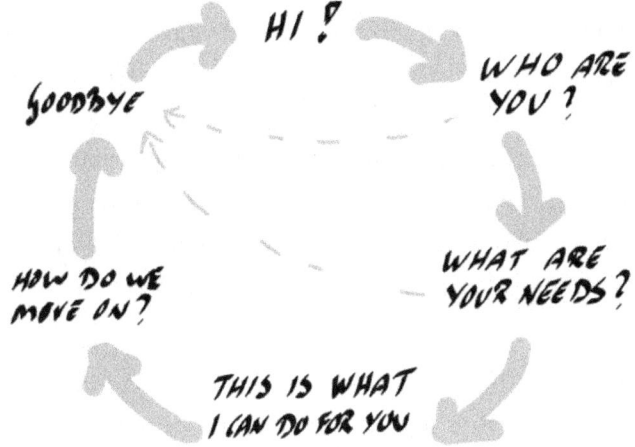

When you know what you want to accomplish, have the right conditions to keep up your energy and feel safe and motivated, you have paved the way to creating really good meetings at the exhibition.

"But how do I make a first contact without being obtrusive?", "What if I stop the wrong person?", "What if the attendee is in a hurry and doesn't have time to listen to me?" These are normal worries for booth staff, even among the most experienced salespersons. Working in a booth is not like an ordinary day at work, you need to learn how to do things, prepare for every step of the way and preferably practice in different situations before entering the exhibition hall. So we have created this model, the Meeting Wheel.

Constantly move in the Meeting Wheel

The Meeting Wheel is about the need to make the first contact, to get to know who you are talking to, to show how you can be useful, to decide how to move forward and eventually how to wrap up the conversation. You, as booth staff, must always be somewhere in the Meeting Wheel while working in the booth. That way, you will always have someone to talk to, you will have more fun and eventually you will contribute to a better result for your organisation.

In the following chapters, we will go through all parts of the Meeting Wheel, with added tips and advice.

Pan for gold

See yourself as a gold prospector at the exhibition. You are standing by a river with your pan and sieve, knowing that somewhere among all the grains of sand there are valuable golden nuggets. But you know that you can't pluck the golden nuggets straight from the river. You must take an entire handful of sand in your pan and swirl it around until you have only gold left in your pan.

This is what you must do at the exhibition. It's no good standing around waiting for the golden nuggets hidden among the mass of visitors to jump into your booth. You have to be active and reach out to a lot of visitors before you find the gold!

The Meeting Wheel Phase 1: Hi!

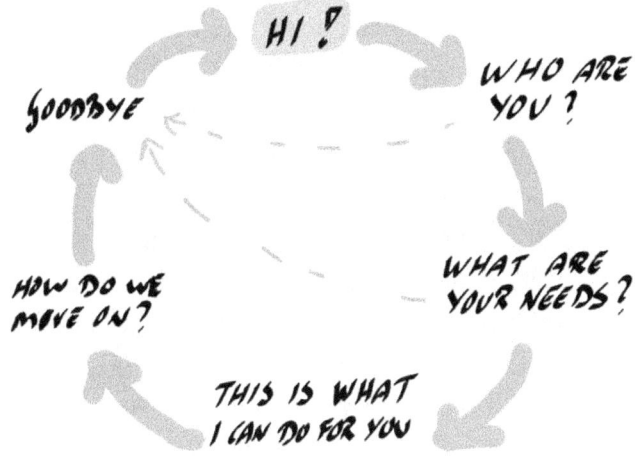

The first step in the Meeting Wheel is the first contact. Sometimes visitors will approach you, but in most cases, you will have to take the first step. You must make a good first impression, be approachable in the booth and, of course, have a great opening line.

Make a good first impression

Look at the squares below. Which one do you prefer, Person A or Person B?

Person A	Person B
Smiling	**Angry**
Well-Dressed	Well-Dressed
Angry	Smiling

90% of people, in average, spontaneously answer A. It is called the 'priming effect' and it means that we put a lot of weight on the first thing we experience. The signals to the emotional centre of our brain work twice as fast as the signals to the analytical part. That means that we make decisions based on our feelings and then we justify them as much as we can. According to the communication experts Patryk and Kasia Wezowski from the Centre for Body Language in Belgium, 70% of first impressions last. Even if we eventually discover that Person A and Person B have the same qualities, only in a different order, our initial impression remains.

Think of how you portray yourself before you even open your mouth.

The same goes for the visitors' first impression of you. Very quickly, they will form an opinion—decide whether they like you. In that short time, they can't possibly judge what you are made of, but they think they can. After forming a quick opinion of you, the next stage is called the 'halo effect'. This means that they will attach qualities to your character, still free of any evidence. If they like you, maybe they will think you are intelligent, helpful, curious, kind and good-looking. If the visitor senses that you are positive and interested, chances are that the first impression is positive as well. This works even better if you deliver a smile and eye contact.

If you want to create a good first impression, send out the right signals:

- Wear clothes which represent your organisation. Exactly what depends on your brand and profile and on the general dress code in the business.
- Neat and clean! If the shirt should be tucked in, make sure that it is. If the clothes should be wrinkle-free, let it be so!
- Look happy and alert. You don't have to pretend to be someone else but be your best self. Be an energetic version of yourself in the booth.
- Look the visitors in the eye and smile with both your mouth and your eyes. Make them feel welcome to your booth.

Be accessible in the booth — do's and don'ts

Your body language doesn't only make a first impression, it also sends out signals to the visitors—telling them whether or not it's okay for them to contact you.

These are common mistakes that make you seem inaccessible to the visitors:

- Standing in the far corner of the booth
- Standing behind a counter
- Crossing your arms
- Leaning on furniture in the booth

- Sitting down
- Looking at or speaking on your phone
- Talking intensively with your colleagues
- Eating
- Twisting your body away from the visitor

Conversely, this is how you signal availability:

- Stand further out in the booth, near the entrance (but not too far out, since that may be perceived as guarding your booth)
- Look up and out at the visitors
- Stand facing out, so that the people passing will see you straight on
- Smile and look happy
- Stand up straight
- Show your wrists
- Open your shoulders, it is a sign of being safe and approachable. Stretch out your arms in a welcoming gesture.

Actively reach out to the visitors

Even if your body language is perfect, there is still a risk that you will be standing there in the booth, all alone, while visitors just pass you by. There you are, with that smile feeling sillier by the minute. Pointless. Valuable time wasted. Boring. You don't have to put up with it…wait no longer than a minute before actively reaching out to a new visitor.

Gather your courage, leave your comfort zone and
reach out to the visitors!

Talk to strangers

Actively contacting the visitors is what most exhibitors find hardest, and that is not strange at all. Ever since we were children, we have been taught not to talk to strangers. Few of us practice talking to unknown people in our everyday life. Those of us who live in a big city have developed the habit of ignoring and 'looking through' our fellow man. And then we are plonked in the booth, expected to do something we have been warned against, never done in normal life, and not been trained to do! So of course, it feels awkward. But we promise you: the more you practice, the easier and more enjoyable it will become!

As a host, it is not pushy at all to make contact

Many exhibitors also avoid making contact because they do not want to be perceived as pushy salesmen. But you are not a vendor in a mall, trying to sell merchandise to everyone passing by. You are an exhibitor at an exhibition. The visitors have come to the exhibition because they are interested in the companies in the market. The exhibitors and the exhibition manager have invited people to the exhibition. You are hosts and the visitors are guests and a good host greets his guests and finds out how he can be of use. If you use the Meeting Wheel as a model, you can never be too pushy.

Ask questions that make the visitor stop

What should you say then to make the first contact, to stop the person you are reaching out to from moving on?

Just saying "Hi!" just gets you a "Hi!" in return. Other ineffectual opening lines are: "Hi, do you want to compete in our booth?", "Hi, do you want to try out our new product?", "Hi, do you want to know more about our new service?", "Hi, can I help you?" All these questions are *closed*, and the visitor can only reply with a Yes or a No. There's a high risk that at least half of them will give you a No and walk on by.

If you instead ask an *open* question, there is a better chance that the visitor will stay. Open questions are those that begin with What, Which, Who, How, Why, When or Where. Open questions incite conversation and give the visitors the opportunity to expand their line of reasoning. Remember that our favourite thing to talk about is ourselves.

Every person's favourite subject: Themselves.

Examples of open questions at the exhibition:

- Hi, where are you from?
- Hi, what do you do for a living?
- Hi, what company are you from?
- Hi, why are you here at the exhibition?
- Hi, what's your interest in this industry?
- Hi, what do you think of our booth?

Other than guaranteeing that the visitor stays and answers your questions, there are two additional advantages to leading with an open question:

- You show interest in the visitor instead of talking about yourself straight away, which is often perceived as nice, empathetic, friendly, open—a good first impression.
- You will get a bit of information about who the visitor is and what angle you might take.

Before the exhibition, come up with at least three opening questions that you feel comfortable in asking. Remember that they should not be answerable with a Yes or a No. Practice with your colleagues, even if it feels silly. You need to know your opening questions!

Why are you here?

For eight years, I was employed at a global company that had booths of about 6000 square meters. There were over a hundred people who worked in the booth and I was one of them. One of my best opening questions to visitors who had climbed aboard our booth was: "Why did you come visit us?" I naturally said this with warmth in my tone. The visitor felt trust and told me. I gained a lot of valuable information that we then could explore further.

Antoni

Who should you reach out to?

Who are you trying to contact? The easiest way is to approach the visitors that are already standing by your booth. Maybe someone has grabbed a pamphlet or been intrigued by an activity in your booth. If there are not any visitors in the booth, it is fine to contact a visitor who is slowly walking past. Just take a step out, smile and ask one of the opening questions that you have prepared.

Challenge your prejudice and your habits

It is human to have preconceptions. But do not miss important contacts just because you are clinging to them. There are natural explanations as to why we are drawn towards people that remind us of ourselves, but it is not professional to let prejudice and old habits decide who we reach out to. A lot of women in male-dominated industries report that they are often ignored by exhibitors, who automatically start talking to their male colleagues instead.

"Look after the old lady"

Lena worked as an assistant in a booth exhibited by a large Swedish company at the Furniture Fair in Stockholm. During the second day of the exhibition, an old lady dressed in a skirt, flat shoes and a long coat stepped into the booth. It was not very busy in the booth and there were several salesmen available, who would have had time to greet the somewhat unusual visitor. The salesmen quickly assessed her and decided that she could not be an interesting visitor and they did not even bother to say Hi.

"Look after the old lady," one of them whispered to Lena, who immediately approached her and welcomed her, and asked what she did for a living. An introduction in English, with a distinct French accent, followed and it turned out that she was the editor of Le Monde, one of the largest daily newspapers in France, and that she wanted to write an article about Swedish design. Thanks to Lena, the company was mentioned in the article. And the salesmen learned an important lesson: to never judge a visitor by their looks.

Åsa

Don't hide behind the activity

Maybe you are having some kind of activity to draw people to your booth? It could be a competition, a challenge, a demo of a product or handing out food or drink. Remember that the main purpose of the activity is to attract people to your booth. And why that? Well, because you will get an opportunity to talk to them.

When you can sense the desperation

At an IT-convention, there were a lot of people in and outside of one particular booth. I became curious and approached the booth. It turned out that an artist was body painting a beautiful woman. When I arrived at the booth, he had just finished. All the visitors then left. I can't remember whose booth it was. I'm guessing that the other visitors couldn't either...

At another exhibition, there was a company that had, a little desperately, set up a video game in one corner of their booth. They managed to draw in some visitors who wanted

to entertain themselves for a while, but the staff had no plan
of what to do. When the game was over, the visitors thanked
them for the moment of relaxation and moved on.

Antoni

So many times we have been invited into a booth by a happy exhibitor to participate in a competition. We compete, leave our answer sheet and get a Thank-you and Goodbye from the exhibitor. In the best cases, we have been briefly pitched at while we were competing, but in the worst, we have barely registered what company we have visited. Win or lose in the competition, a visit like that will not leave a lasting impression.

Take the opportunity instead to talk to the visitors that have been attracted to your booth by your various activities. Start the conversation by asking an open question and then continue by being curious and interested!

"There's no such thing as a free cinnamon bun"

A customer of mine was once part of the most popular booth at the exhibition because of their activity. They were handing out coffee and cinnamon buns. So far, so good. But after the first day, they realised that they had not gathered any contacts they could follow up on.

It was easy to grasp why the salesmen had failed so profoundly with their encounters—they had not had the time to talk to the visitors because they had been busy serving coffee instead. "No matter how much we served, the queue was always long," one of the salesmen complained.

The next day, the sales manager presented a new strategy. Only the two booth hosts, who were standing behind the counter, were to serve coffee. The sales team could only fetch coffee if they were engaged in a conversation with a visitor. The queue to the coffee counter was an excellent opportunity for the sales team to reach out and engage in a meeting. The visitors were just standing there waiting.

In the two days that followed, the sales team got more leads than they were dreaming of. All visitors got their coffee and cinnamon bun, it just took a little longer and they got to have a nice conversation with one of the sales persons of the company. There's no such thing as a free cinnamon bun.

Åsa

"Too many" visitors?

Sometimes the rather pleasant problem arises that so many visitors are interested in contacting you in the booth, that some have to stand and wait their turn. Have the following in mind:

- Always indicate, with a smile and with eye contact, that you have noticed the visitor who is waiting.
- Ask for permission to put your current conversation on hold to tell the waiting visitor how long it will take before you get to him.
- If feasible, invite the waiting visitor into your current conversation.

- Make sure that the waiting visitor is occupied: by participating in a competition, having a cup of coffee, trying out a product, etc.

A summary of the Meeting Wheel phase 1

First contact

- The first impression is important: be clean and neat, look happy and wear clothes that represent the company you work for.
- Show the visitors that you are available and interested in making contact by smiling, standing straight and turning towards them.
- Never do any of the following in your booth: talk into or look at your phone, sit down, lean against furniture, stand with your arms crossed or turn your back to the visitors.
- Prepare at least three open questions which you can ask to start a conversation. They should not be Yes- or No-questions.
- OK, it can be intimidating to reach out to strangers, but you still have to do it. The more you practice stepping out of your comfort zone, the easier it will become.
- Use competitions or other activities to start conversations and create meetings, not as an excuse to get out of talking.

The Meeting Wheel Phase 2: Who Are You?

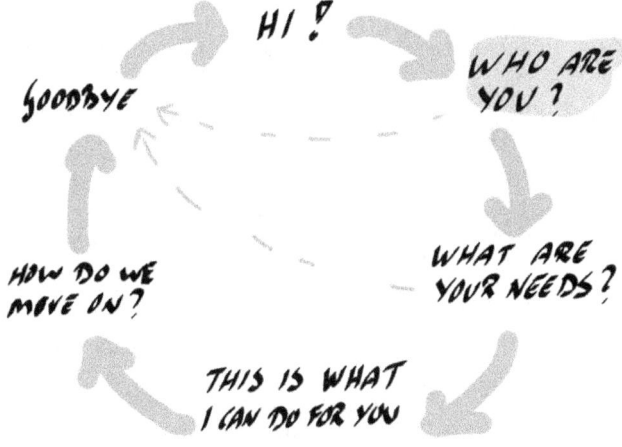

In this step, you make the first sorting out, to quickly assess if this person belongs to your target group or not. Needless to say, all people could be interesting to talk to but at the exhibition it is important that you determine, as quickly as possible, if the two of you would gain anything from continuing with your conversation. Spend the proper amount of time on the right people—without being rude to those who only merit a few moments.

Think about what type of visitors that will come to the next exhibition. Use this checklist and add your own categories:

- Clients
- Potential clients
- Competitors
- Partners
- Possible partners
- Media
- Colleagues
- Influencers, such as regulators, politicians, lobbyists, the customers of clients etc.
- Gatherers of pens
- Other exhibitors
- The general public (even at trade shows)
- Schoolchildren

You will quite likely meet people from all these categories—many of them "not our target group". So, as part of the booth staff, you must quickly assess who you are talking to.

Find out if the visitor belongs to your target group

Prepare great questions in advance, to quickly determine if the person you are talking to is worth your time and energy. At most trade shows the visitors have name tags where you can see information about what company, city and visitor category they belong to…but do not always assume that the information is accurate. It is the visitors themselves who have registered and sometimes they fill in the wrong box. Besides, the text could be too small to read or the name tag hanging round the visitor's neck could be facing the wrong way. It is better to talk to the visitor and ask questions without relying too much on the name tag. Otherwise, you might easily miss interesting and important people.

When you know who, you can ask what.

Do not settle for asking just one question. Be curious and ask follow-up questions! Suppose that you are selling IT-systems and you ask a visitor what company she works at and she answers that she is an employee at the company that is your dream customer. You become happy and eager and quickly start talking about technical qualities that you know by heart, but you notice that the customer becomes restless and is checking her watch. What you have forgotten to check is what department she is from. Maybe you have the financial manager in front of you. Maybe it is someone from HR or the sales team. Every function has its own agenda, priorities and interests, so keep on asking to get a complete picture of the visitor.

"It is not that heavy!"

In the year 2000, I was standing on the floor of CeBIT, the world's biggest electronic exhibition. I was showcasing new and innovative location services on what was the first "smartphone" from Ericsson. This was long before all phones had a GPS. On one occasion, a man approached me and seemed interested. I did not ask any questions but kept going at high speed about our new location based service. In a mix of German and English, I tried to get him interested. After ten minutes, I had nothing more to say.

Then he held out his hand. I handed over the phone and was expecting him to look at the app. He did not. He only weighed it in his hand and said: "Ah, it's not that heavy." Then he smiled, handed me the phone back and left...

Antoni

What questions to ask at the beginning of a conversation depends on what organisation you represent and what your target group is. Here are some examples:

If you are representing a university that exhibits at a student show:

- What program are you in?
- What subjects are you interested in?
- What would you like to work with in the future?

If you are selling ventilation systems at a construction show:

- What company are you from?
- What is your position?
- Have you purchased products from us before?
- How important is it to your clients to have well-functioning ventilation?

If you are a company dealing with care and welfare at a social work show:

- What do you work with?
- What county do you work for?
- What challenges do you have?

What do you need to know about the visitors to quickly determine if they belong to one of your target groups or not?

The elevator pitch

It is common for the visitor to rather quickly ask the questions: "What does your company sell?" and "Why are you at the exhibition?" Maybe they are reaching out to you and this is a step in their Meeting Wheel!

Do not be too hasty and risk starting to sell before you know who you are talking to. You can give a very brief presentation of what you have to offer. Then you can assess in what way it would be interesting to the person you are talking to and return to asking questions to get a picture of their needs—your opportunities.

This type of presentation is often called the "elevator pitch" (which originates from the US, where you can imagine that you have, say, 20 floors to catch the interest of the CEO who happens to be in the same elevator car as you). You have about 10 seconds to half a minute.

The best thing you can do is to prepare and test different pitches before coming to the exhibition. Write them down in spoken language so that they sound natural to you as you read them.

Practice elevator pitches so that they come out naturally when you work in the booth.

Common parts are:

- I am/we are
- We are offering
- This is how most people benefit from us
- In what way would this be interesting to you?

Or

- Most people that come to me/us are...
- They want value X/ to solve problem Y
- We have a solution that will help them...
- In what way would this be of interest to you?

With the last question of the elevator pitch, you return the ball to the visitor's court. You are the one steering the conversation while showing that you want to get to know the person in front of you.

The one who asks the question steers the conversation.

When the visitor isn't part of your target group

If you realise that the person you are talking to does not belong to your target group, you need to end the conversation as quickly as possible. It could turn out to be another exhibitor, a rival, someone who doesn't want to buy your products or maybe someone who works for the exhibition.

You should always aim to be nice and friendly, but use your time wisely. Helping everyone feel welcome is a very good start, but you shouldn't spend too much time on the visitors who are not interested. You will be doing them a favour by letting them move on. In such cases, you take a shortcut to phase #6 in the Meeting Wheel: Goodbye.

It does not have to be complicated to end a conversation with the visitors who are not in your target group. It may be enough to say: "Well, I won't occupy your time any longer, have a good day at the exhibition." And maybe you are such a good host that you will guide them to the part of the exhibition where they will find something useful.

You will also encounter "time thieves"—people who are not in your target group, but who love to stop and chat...for a bit too long. Remember that everyone who visits an exhibition has something to do with the exhibition or the industry—you can gain interesting information, knowledge or insights from them and they might all be your ambassadors even if they do not fall into the category "potential customer". Having a chat with someone can be nice, but it is important that you realise when it is time to take control, end the conversation and say goodbye.

Then there are always competitors who want to come in for a chat. They are, most likely, not part of your target group either. We suggest that you are as open as possible and ask them questions, lots of them. Naturally, you will not expose any trade secrets, but what you showcase at the exhibition should be able to stand up to your competitors' scrutiny. And who knows?…maybe someday the competitor might be your colleague.

The time I got turned down 100 times

When I visited an exhibit about the exhibition industry in Miami, I was the unwanted visitor. None of the American companies were interested in having me as a customer and I was not interested in buying anything. I was just there to check out the exhibition.

The exhibitors were very good at approaching me as I walked by. They asked a few engaging questions that quickly had them deciding that I was not a potential customer. "Well, nice talking to you, have a safe trip back to Sweden," they said and then handed me a giveaway before saying goodbye. They had ended the conversation in a nice way, just a minute after they had approached me. I was still content—I did not want to unnecessarily steal their time! And I have never been given as many giveaways (or perhaps it was go-aways?) as I had at that exhibition.

Åsa

A summary of the Meeting Wheel phase 2

Who are you?

- Before the exhibition: Prepare the questions you should ask to find out if the visitor belongs to any of your target groups or not.
- Find out who the visitor is before you give him your pitch.

- Be curious and ask follow-up questions but be effective. You must quickly build a complete picture of the visitor.
- Challenge your own prejudices and habits so as not to miss interesting visitors.
- If you deduce that the person you are talking to does not belong to your target group, move directly to phase #6 of the Meeting Wheel: Goodbye!

The Meeting Wheel Phase 3:
What Are Your Needs?

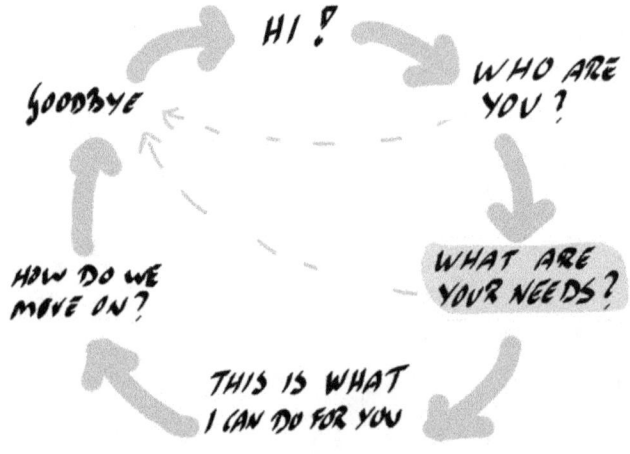

Now, you are standing there with a visitor who does belong to one of your target groups. It could be a potential customer, a current customer or just someone who might be valuable, like a journalist or a potential business partner. But just because they belong to your target group does not necessarily mean that they need your help. Now is the time to keep probing, to determine their need—and letting those without any need of your products or services take the shortcut to the Goodbye-phase in the Meeting Wheel.

Prepare in-depth questions. What do you need to know about the visitor to be able to present a solution?

Ask more questions to get a deeper understanding

It is very rare for a visitor to step forward and give you the entire picture: what need they have, what they are after and what can solve the issue, what budget they have and when they need it done. Therefore, it is up to you to keep asking questions. The questions should lead to answers that tell you more about the visitor's:

- General business situation
- Problems, wishes
- Consequences if the problem is not solved or the wish not satisfied
- Value of solving the problems, satisfying the wishes
- Range of possibilities

Both open and closed questions work — as long as they are relevant

Keep asking open questions, but in this situation you can also use closed questions as you check facts or check that you have understood the visitor, for example: "As I understand it, you are planning to make a large purchase next fall?", "Are you the one who is in charge of purchases?", "Are you happy with your current provider?"

There are still dangers with closed questions, however. You need to do all the thinking yourself and the customer might feel that you are mostly checking off a list. About 60 percent of the times you ask a closed question, the counterpart will give a more elaborate answer, so you are not in trouble, but be prepared to ask relevant supplementary questions. Open questions are better if you are looking for information. They are shorter and 90 percent of the time they will be answered with more than one word.

The shorter the question, the longer the answer.
The longer the question, the shorter the answer.

Be like a four-year-old, ask why!

Way too often, we accept the first answer that comes out of the visitor's mouth and we forget to ask follow-up questions that will help explain, deepen and clarify. We always need to understand better.

Sometimes our own experience makes us feel as if we can read minds, but we can't. By following up with variations of Why-questions, you will quickly get to the true reason behind the need and consequently, the true value. With your questions, you are contributing to your own understanding—and sometimes even the visitors will understand their own needs better.

Some people, however, find the "why-question" offensive and we partly agree with this. If you have not established a good connection with the visitor, too many *why's* can be provocative. Your entire tone while asking can make the visitor feel either aggrieved or very important.

Strategic questions

If you want to be more than just a standard, off-the-peg provider, you need to ask strategic questions—find out what the visitor's goals are and why. The answers expose the true value of any deal that might eventually come about.

Examples of questions are:

- What do you hope to gain from the product/solution/service?
- What is your goal?
- What do you need to be able to solve the problem?
- What criteria are valued in your department?
- If you do not solve the problem, what happens next?
- If you do solve the problem, what do you gain?
- What is important to you? Why?

Strategic questions trigger the right half of the visitor's brain, where the visions are. If you can make the visitor describe the goal, it is much easier for you to understand what the visitor wants. Moreover, if you understand why the visitor has that goal, you will be right at the heart of his need. With that knowledge you will add the right value. Needless to say, it does not have do with the visitor's grand business visions, but rather more ground-level problems that he wants to solve…what he wants instead of what he's got right now.

In other words: If you can understand the visitor's goal and purpose within the range of what you have to offer, and if you show the visitor that you understand, half the work is done. At least.

What do you want to achieve? Why?

If it's going well, you can follow up on their answers and find out how much they really want to solve their problem. If you help them become aware of that, they will, potentially, see you as the key to a better tomorrow.

Leading questions

Moving on, we come to something called *leading* questions. These questions make the visitors focus their answers on an area you have decided you want to know more about, and where you have something to offer. For example: "If we enter area X, what are your thoughts around that possibility?" Leading questions can be useful when the visitor shoots off in every possible direction and needs guidance and help to focus. But be careful not to appear an insensitive or manipulative person who just wants to make a sale.

Observe the way others do need-analyses and discuss it with your colleagues. How do they express themselves? Do they ask long questions or short ones? Open or closed? Leading? How do you do it?

Listen to understand!

It is good to ask relevant questions, but the visitor should not feel as if you are checking off a list. The visitor will quickly notice if you are truly interested—or if you are faking it. In this phase, as the visitor is clearly interested in talking to you, it is more important to be interested than interesting.

Be interested and you will become interesting.

Having someone really listen to you is a luxury we seldom enjoy. You can give the visitor that feeling if you are present and curious. There are different levels of listening, where you are more or less active:

1. Level 1. You listen only to respond. This is the shallowest kind of listening, barely better than just hearing (= registering sound waves). All you do is listen for keywords that immediately make you want to end the chatter coming from the visitor, so that you can talk instead. Imagine this: you have a demo of an CRM-system that you know can help customers with different needs, such as efficiency, increased productivity, happier customers and so on. If the visitor says, "Efficiency is important..." and you cut the person off to tell him everything your CRM-system can do to increase efficiency in the sales force, then maybe you have missed what was coming: "...but right now we are putting all our focus on finding new customers".

2. Listening at level 1 is like eating dinner with a group of friends. One person talks about something that has happened and the next trumps that story with one of his own, and so it goes on. Everyone is on transmit. No one is paying attention. It can be pleasant, but it will be more like entertainment than an actual exchange.

3. Level 2. You listen to understand the message. This is more like it. You repress your need to immediately

formulate a response. You let the visitors finish their sentences. You build trust and ask in-depth follow-up questions. You listen to what is said and what is not. At this level, you begin to ask follow-up questions like: "What do you mean by…?", "Where did that lead?"

By echoing something that the customer says, you can also gain a lot of information. For example:

Visitor: It is important to us that we get the course literature on time.

Seller: On time…?

Visitor: Yes, or we'll have trouble distributing them to the classes. The day before would suffice, but no later than that.

Now the seller knows: the day before will be good enough.

Listening at level 2 is like having a confidential conversation with someone who is giving you enough time to develop your thoughts.

Level 3: You listen to understand the person. When you have someone in front of you, you can choose to both listen to what is being said and observe how it is being said. Is the visitor leaning forward and nodding? Are they looking down at the floor, answering evasively? Do they start waving their arms and raising their voice? All the signals that are being sent to you can be interpreted. Not with one hundred percent accuracy, but it all helps.

At this level you are better at discovering what is being left unsaid. Often there are plenty of parameters behind a decision, other than just looking at the qualities of your product. Sometimes there are political games that you are unaware of. Sometimes things happen in your visitor's private life that affect their behaviour. Even if you can't work out the details, you can sense the mood and signal that you are doing your best to understand and that goes a long way. Having someone who listens at level 3 is like having your own personal coach, who focuses most on your agenda.

Asking is silver. Listening is gold.

Practice at home, at work and when you meet your neighbour on the bus. Your ability to listen at levels 2 and 3 will help you better understand the visitor's need, which means you can soon show what help you have to offer.

A summary of the Meeting Wheel phase 3

What are your needs?

- Before the exhibition: prepare by deciding what you need to know about any given visitor before you can offer a solution.
- Try to get a deeper understanding of what the visitor needs.
- Ask the question "Why" as often as you can.
- Let the visitor talk.
- Listen to understand, not just to respond.

The Meeting Wheel Phase 4:
This Is What I Can Do for You

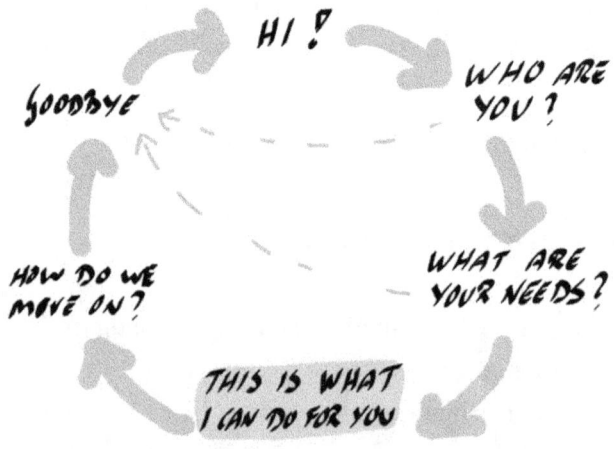

Now we come to the delicate part of the conversation, when it is time for you to talk about how you can be of use to the person you are talking to. What you want to avoid is pressing the "play button" and then monotonously droning on about yourself, your organisation and your products. You want to give the visitor interesting and relevant information which can help them decide.

There are some challenges:

- The visitor is in a hurry, because there are more exhibitors to visit.
- The visitor does not take in everything you say—an exhibition is a noisy, distracting venue.
- The visitor only gets interested if what you are saying is relevant to him or her, which requires that you have done your exploration in the previous phase.

You need to give the visitor a clear picture of how you can help them in their challenges in an efficient and effective way.

Help each other

Do we have to be able to answer any question? This is a common concern raised at our courses for booth staff. The bigger the company you represent, the harder it will be to know all the products, services and solutions. If that's your situation, it is important that the project leader mans the booth with experts in various areas.

If you realise that a visitor needs information that you yourself do not have, make sure to direct him to a person in your booth that does have the knowledge. Guide the visitor to that person and quickly sum up the conversation so far.

If there is no expert on that particular area in your booth, ask for the visitor's contact information so that he can be contacted by the right person.

Make sure to find out who in your booth is a
specialist in what area.

Tailor your message with DISC

Not every visitor will respond to the same presentation, so you must be prepared to present your offer in different ways.

Some people want to go into details, some want the history behind your products, some like to take things slowly, some are in a hurry. The analysing tool Extended DISC divides people's communication styles into red, yellow, green and blue. It can be useful to apply this model to your visitor: what colour might they be?

If you meet a person with a red communication style (the **D** in DISC stands for Dominant), you may notice impatience, lack of interest in the small details, and a direct and urgent demand to know if your solution can help him or not. (There are slightly more men than women in this category, according to a study from 2011.) The key words of the red profile are *Results* and *Winning*.

In someone who has a yellow communication profile (the **I** in DISC stands for Influence), you will notice that she is outspoken, easy-going and often inspiring. She (slightly more women than men) can share a story or two about herself. The key words of the yellow profile are *Inspiration* and *Relation*.

Then there's the individual with a lot of green (the **S** in DISC stands for Stability)—most likely a very caring person (slightly more women than men), who puts the prosperity of others first. The key words of a green profile are *Loyalty* and *Accuracy*.

The fourth category is the blue one (the **C** in DISC stands for Competence). This guy (slightly more men than women) is interested in facts and analyses, and may find feelings less important, even disturbing, when decisions are to be made. The key words of the blue profile are *Facts* and *Quality*.

A word of warning: What you have just read is only a model. It's a rough generalisation—and should not be taken literally. We humans are multifaceted and complicated. Use this as an indication if you want to, and if you want to learn more, then make a personal profile of yourself and the others in the booth. Having a model like this in your armoury can help you understand both yourself and the others better.

What kind of personality are you closest to?
How can you act to make sure that you are on the same
wavelength as this or that visitor?

From awareness to action

AIDA is a model showing the stages any person goes through before deciding on a purchase: Attention, Interest, Desire and Action. Where on the scale is the visitor you are talking to?

Awareness

If the visitor knows nothing about what you are offering, give him a clear idea. Get him to focus on you and the coming conversation amid all the distractions of the exhibition out there.

Interest

If the visitor is aware of what you've got, but still has no feelings about it, it's time to ask relevant questions: what matters to them and why? Once that is established, you immediately become more interesting—as will your offering. Relevant, attractive, appealing.

Desire

Here you have a visitor who likes what you are saying and what you are offering. It could be because of you as a person (people buy from people) or because they have grasped the use and value and are ready for the next step.

Action

The visitor now has all the information he needs and is keen on buying. Don't drone on about features, benefits, market trends…This is the moment to show them how they can actually make a purchase, or get a follow-up meeting.

It might take time to go through these four phases—even a series of meetings. Sometimes it happens fast—just like that! So be attentive, responsive and flexible.

How storytelling will help you

There is probably no technique, no tool, no key as effective as a well-told story. It's borderline magic. If you can plant experiences in the visitor's head and heart, they will remember more than if you feed them only facts. Stories win over logic. That's how the brain works.

Your brain has several important parts. There's the neocortex: the logical brain—like an analytical doorman. If you only hear information based on facts, the doorman can simply shut it out, especially if those facts contradict what you already know and believe. But if you're offered a story, it will reach the emotional part of the brain, the limbic system. Here the brain can enjoy, be touched, be appalled and rejoice. Then, if the story is seasoned with useful facts, the analyst will once again get involved and rationalise the emotional decisions.

Stories boost the value by creating emotional connections to what you are selling, and to you as a representative of the organisation. Use stories, examples, parables and metaphors while talking about your product, service or solution. How did other customers do it? What will the future look like if the visitor chooses you, or rejects you?

The basic ingredients of a story are a time, a place, a person, a problem and a solution.

Answer the following questions and you will have a good platform:

- When did it happen? Recently? Summer, winter, night or day?
- Where did it happen? Here or far away?
- Who did it happen to? An individual, a group...maybe another customer?
- What happened? What challenge, difficulty or possibility appeared?
- What was the solution? In what way did the customer or you solve the problem or seize the opportunity?

Remember that we humans remember things by using our five senses, so if you want the stories to truly make an impression, describe each step with one of our five senses. The visitor can then see what you see, hear what you hear, feel what you feel and so forth.

Prepare your stories in advance and share them with your colleagues. What success stories do you know? A story doesn't have to be true, factual, real; it can just be a realistic, credible illustration.

Do you sell features or true value?

Everyone who visits you in your booth invests the single most important resource they have: their time. Everyone wants the answer to the same question: How might this benefit me/us? What's in it for me? If you explain what you have to offer and the customer answers with a "And...?" or a "So what?", then you have not shown them how you can be of use to them.

Millions of screwdrivers are sold around the world every year. But it's not the actual screwdriver we want! We want a pretty house, we want to hang a picture, build a balcony or put together a piece of furniture. We're not buying the screwdriver for the sake of the product, but for what we can do with it and the value it brings.

Practice answering the question "What's in it for me?"
asked by different categories of visitors.

If you have created a good atmosphere, asked the right questions and listened to the answers, then this phase is very simple. You only need to focus on what the visitor has said. You only offer what is of value. Nothing else. You don't even have to get into how you will solve the problem, at least not in detail (we'll get to why later). The visitor will have gone from an awareness to an increased interest—ready to move forward with you.

If you ignore the need-analysis and choose the "hair dryer" method instead (talking so much you blow the visitor's hair back), you will have no idea of what is important to the visitor. It will lead to you presenting argument after argument and saying a silent prayer that something will stick. And that leads to two things: 1. The visitor realises that you are not interested, 2. The visitor will think your offer is uninteresting or unnecessarily expensive because you are only talking about qualities that the visitor does not need.

A simple way of structuring your pitch is the FBI-model. It stands for:

- Features: your product's or service's weight, length, width, speed, precision, quality, etc.
- Benefits: what is special about the features—faster, smarter, lighter, higher, cheaper…
- Incentive: what it would mean for the visitor to use your service or product ("What's in it for me?"). This value for the customer must be grounded in the answers you got to the beautiful questions you've asked. It could be higher status, more money, greater security, increased profitability, happier customers…

Example of the FBI-model

Imagine that you are at an airport, on your way home.

A person comes up to you and says: "I have a cab. I can drive you home in 20 minutes rather than an hour on the bus. It'll be 75 dollars."

The cab is the feature. It is a way to get home.

20 minutes is the advantage. It is quicker than taking the bus.

Imagine the cabdriver asks if you are interested in football and you say Yes. Then he can also give you an incentive, which will be the strongest selling-point: "If you take a cab home, you can make it home to see the entire game tonight!"

A 20-minute cab ride might not be worth 75 dollars, but to make it home on time to see the game is what makes you accept the offer.

We will buy if the perceived value is greater than the cost. Otherwise, we will not.

What you say and how you say it

How will you present your product, service or idea? Mere information will soon make you boring. Speaking only with high energy but no substance will make you a clown. Your choice of words, your tone of voice and your body language are truly important in convincing the visitor that what you are saying is right and true.

Words matter. The right word in the right place can work wonders. A good tip is to listen to the visitor and find their favourite word. Use that word when you talk. You can also use clearly negative words for the problems and challenges and clearly positive words for the solution. That will make your solution even more attractive.

Avoid words like:

- A little
- In principal
- Maybe
- Possibly
- I thought that I just…
- One can…

They lower the value of your communication if they are used often. It will seem as if you don't believe in what you are representing.

Be careful not to use technical terms and abbreviations, unless you are certain that the visitor knows them.

Another great tip is to find the core of your offer and use fewer, better words so that the visitor will not leave drenched in verbiage.

The voice can wake you up or lull you to sleep. Your voice is important because it creates and mirrors emotions. It portrays just how much of your soul you are putting into the conversation.

A few examples:

- If you emphasise certain words, it will be easier to understand that they were important.
- If you vary in volume, tempo and tone, you show that you are committed.
- If you pause after having made an important point, you give the visitor the joy of catching up with your reasoning.
- If you pause before making an important point, you give the visitor a feeling of sharing a secret. Especially if you also lower your voice while telling him.
- If you have a monotonous voice, people will find you uncommitted and boring.

But how will you know what you sound like? One simple way is to record a few customer conversations, then pick up clues about how you are using your voice. Maybe you interrupt too often or use verbalised pauses (*like, um, eh, you know, well...*) a bit too often?

Your body language strengthens or weakens your message. Earlier in the book, we talked about the importance of body language to make a good first impression and to not scare off the visitors. Here we bring it up again. Body language counts in making a visitor believe in what you are saying. If you say something to the visitor, but your body language says something else, the visitor will put trust in what he is seeing rather than what he is hearing.

How you say something weighs heavier
than what you are saying.

You don't have to be an actor. You don't have to be anything but yourself. But—Be your *best* self. If you think that your products are good, show it with happy eyes and a happy mouth and a proud posture. Below are some examples of how your body language can strengthen or weaken your arguments:

You say: "We can lower the cost by 20 percent."

You show it with your body by lowering your hand.

You say: "We will work on your challenges together."

You show it with your body by holding your arms in an including circle.

You say: "Our vision for the future is…"

You show it with your body by pretending to hold a crystal ball in your hand. Or move your hand from your body, creating an imaginary timeline as, for a moment, your gaze travels through the room.

You say: "We never do business with companies that don't follow the environmental laws."

You show it with your body by shaking your head and putting on a serious face.

All these gestures, and many more, enhance your message. Moreover, the words will flow more easily because hand gestures are in the same place in the brain as speech. If you lose track you can literally "crank" yourself up again.

Show with your body that you are listening and that you understand. Focus on the visitor and show with your facial expressions that you are interested in what he has to tell you.

- Look the visitor in the eye. An optimal level in a conversation is 60-70 percent. More than that and you are staring—not good. If your gaze wanders, you will appear nervous. If you look away too much, you will seem arrogant. Note that in some cultures it is impolite to look people in the eyes, especially for too long.
- Nod if you agree.
- Lean forward a little to show that you are interested.
- Show empathy for the visitor. If they are telling you something troubling, show that you sympathise with them.

- People tend to raise their eyebrows, when their eyes are taking in too much information. Find a balance. If you exaggerate, the visitor will be scared of you. If your eyes are more closed—narrowed—you may come across as uninterested.

Match the person you are talking to

This is another something that you can prepare and practice before you go to the exhibition. People who want to be liked, and who are genuinely curious about other people, happily adapt their actions and expressions.

If I am like you, I tend to like you.
If I am not like you, I tend to not like you.

Every dialog is like a dance: one leads, one follows, with the roles switching sometimes. As you meet a visitor the same thing will happen. As you lead, he takes stuff in. Then he delivers information and you follow at his rhythm. Getting on the same wavelength…finding the right chemistry…going with the flow. Both open to sharing ideas.

There are three ways you can match your visitor:

1. With the words you speak. Use the same words that the visitor is using, especially if you notice that they do have some favourite words. You show respect if you use them, rather than approximate synonyms, which might have other connotations and associations.

2. With your body language—cautiously imitating the visitor. If the visitor uses his hands and arms a lot,

while you are standing stiffly, it will not be a warm engagement. And if the visitor takes a step back during the conversation, don't pursue him. It might be that he prefers to keep his distance rather than getting in close and intimate.

3. With your tempo. We often fall into the trap of talking faster when the visitor is talking s-l-o-w-l-y. It feels wrong—like you jumping forward while the visitor backs away.

You don't have to mirror the visitor in everything. For example, if the visitor marches towards you with a scowl and his arms folded, it's better for you to maintain an open, friendly and inviting posture, in the hopes that he might mirror you instead.

Look around on the bus, tram or subway—observe a bunch of teenagers who are friends. Notice what they have in common—clothes, words, tone of voice. You can see they belong together. See if you can spot a group leader. Are all the others copying his or her style?

A summary of the Meeting Wheel phase 4

What can I do for you?

- Before the exhibition, prepare yourself by practicing a set of presentations, varying according to the background of the visitor (including personality) and where in the process of buying he is as you speak.
- Use stories to support facts—memorable, persuasive.

- Sell the values, not the obvious features, and not even general advantages
- Remember that choice of words, tone of voice and body language are at least as important as the information you give to the visitor.

The Meeting Wheel Phase 5:
How Do We Move On?

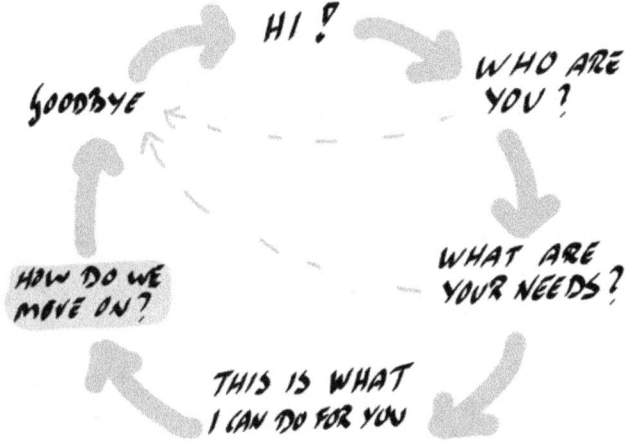

You are at the closing stage of the conversation. You have collected the information that you needed to show and tell how you might be of use. Before you wrap it up, you need to:

- ask tactical questions
- adapt to your visitor's decision-making style
- reach a decision with the visitor

- take notes of the decisions that you and the visitor make

Tactical questions

It is important, even at this late stage, in this part of the conversation, to sense the mood and ask questions. The questions will probably be about practical matters, like:

- What do we do now?
- Who should be involved?
- Who does what?
- How will we deliver?
- When do you need the product/service?
- When can you make a decision?
- Shall we meet and discuss this further?

Adapt to the decision-making style of the visitor

To ask the right kind of questions, or to propose a way forward, you will benefit from learning a thing or two about different decision styles that your visitors might have.

By observing how a person expresses herself, you gain clues as to what decision-making style she uses. Here are the three most common ones:

"I'm in charge"

People using this style make decisions because they have the mandate, coupled with general impatience—often expressed in a 'gut feeling' that tells them Stop or Go. From receiving key information to pressing the button might be a matter of moments. These people often signal that they are the boss, or overall responsible, or in charge of the budget.

What to do? Ask What and How.

They would rather not have you dictating their future. No, they have the mandate and they want to make the decisions themselves. Examples of questions:

- How do you want to move on?
- What is your next step?
- What is your decision?

"I need more time"

People like this want more time—maybe for analysing and comparing, or maybe because they themselves do not have the mandate and have to check with others to get a decision from them, or maybe because they just don't like to feel rushed. They usually send a signal ("Let me think it over", "I'll get back to you on that", "There's no rush"). In some cases, it's just a polite way of saying No, but that shouldn't be happening at this late stage of the discussion.

What to do? Ask Why and When.

You need to make sure that the visitor doesn't sleep too long. Ask questions like:

- What will you use the extra time for?
- How can we help you until then?
- When will we meet again?

"Reassure me once more..."

These people come across as insecure and hesitant, even though you clearly have something that could be of use to them. They need someone who can calmly and safely repeat the needs, opportunities and advantages, to guide them on their way to a safe decision. If they feel pressured or backed into a corner, they might just give you a No Thanks, and miss the opportunity you are offering.

What to do? Summarise and propose.

- Repackage your discussion, step by step, and have them accept—Yes...Yes...—Then suggest a natural path forward.
- If they still are backing away, there is something in your proposal that is making them hesitate. A hidden agenda? Explore further, if you can do it without scaring them off.

It isn't always easy to know exactly which decision-making style the visitor has, but try to be perceptive—use your intuition to increase the chances of a positive decision and a close.

Make a decision about further contact

At most B2B-exhibitions, the real selling happens after the trade show. You can look at the meeting at the exhibition as a first date, and the time to get down on one knee to propose comes later. Your aim as an exhibitor is to get a second date. Depending on where the visitor is in the AIDA process, described earlier in this book, there are different ways to move forward in the buying process:

- The visitor leaves a note of interest to order
- You schedule a meeting after the trade show
- You agree to call each other later
- You agree to catch up via email
- You promise to send more information via email
- You add the visitor's address to your newsletter subscription

There will be visitors who will decline further contact. Even though it feels annoying, you as an exhibitor must respect that. The second-date decision needs to be mutual!

Register all leads

Write down what you have agreed upon, with whom and why. Even better if you do it during the conversation, together with the visitor, so you can make a last check and avoid mistakes. There are systems that manage leads, with devices that scan visitor badges, allowing you to add information about what you have decided. If you have one like that, make sure to learn how the system works. If not, you will get a long way with business cards, paper and pen. The most important thing is that the information be registered somehow.

There are several reasons to be meticulous at this point:

- The next conversation you have with the visitor, after the exhibition, will be much easier if you remember what you talked about. The brain is an odd machine that erases, twists and generalises information. Remembering every detail of a conversation at an exhibition is impossible for most of us.
- When the subsequent work is being done by colleagues who were not at the exhibition, they have a much better chance of following up properly if they know all about the conversation in the booth. Salesmen hate conducting a follow-up conversation that's pretty much small talk.

What can you promise during the exhibition?
How will you document what you have promised?

We realise that it may be hard to find the time to take thorough notes, especially if there are many visitors calling for your attention. But if you do not, you will lose business, have discontented customers and face irritated colleagues after the exhibition.

The booth staff who documents the leads best wins the prize of "Most Helpful After-Event Booth Staff".

Some ways are better than others

During the years that I have trained booth staff, handling leads has been a constant challenge. If the right information is missing, there will be no lead and the visitor may end up disappointed because you did not follow up. I have seen versions like: "Someone from company X was here. They were interested." That information is not valuable.

Some people collect business cards. Others takes notes in a notebook. At one occasion, all leads were documented by the staff calling and recording their leads to an answering machine, which was sent off to a company in India. The day after, the staff had the information in their mailbox. It worked surprisingly well. Other than that, it is mostly common at bigger exhibitions today that the staff scans the visitor's badge and writes additional facts on a tablet, for example.

Antoni

A summary of the Meeting Wheel phase 5

What is the next step?

- As far as possible, adapt to the decision-making style of the visitor.
- If you are not selling products off the shelf, the most important thing is to get a second date with the visitor.
- It's important that you make the decision about the next step together with the visitor.

- Document everything you promise and agree upon, so that you will not fail to follow up on important contacts after the exhibition!

The Meeting Wheel Phase 6: Goodbye!

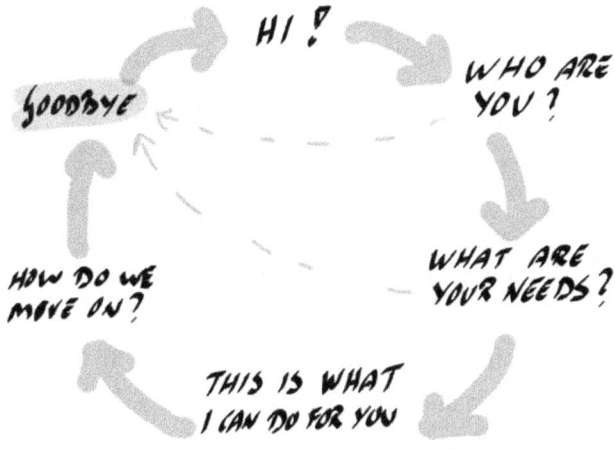

How long should a conversation be?

How long should a conversation be? Answer: Short.

Once you have decided on some kind of contact after the exhibition, you should begin to round off the conversation. As an exhibitor, you might feel like holding on to the eager and curious visitors—a kind of ego-boost. But if you go on with the chat after their interest has peaked, their energy—and their attention—will slip away. They probably have other booths and stands to visit, so don't lay further claim on their time.

Have you made a decision on further contact?
Make sure to end the conversation nicely.

See to it that the visitor has more energy after
your meeting than before.

Goodbye!

The more visitors you find time to talk to, the more possibilities will open up for valuable contacts. It is impolite to say "Goodbye" straight off. You need to end the conversation nicely—with good manners. There are several ways to do that:

- Show your appreciation for the conversation, thank them and indicate that it is now over.
- Ask if there is anything you can do to help. That question alone can generate new possibilities.
- If there is time, sell more or create new areas of interest. Many customers are at their most receptive once they have made up their mind, once their momentum is up. This is also the perfect opportunity to introduce a colleague of yours to them, so that they can get your VIP treatment instead of visiting your competitors.
- Do not babble any longer about your plus-points. Of course there is benefit in sending the customer away

feeling the right decision has been made, but draw the line at needless repetition!

- If you have a giveaway, this is the time to give it away.

The most important thing is that the visitor feels that it will be fun to hear from you again.

Join some colleagues and practice different
ways to end conversations in the booth.
Give each other different scenarios.

A summary of the Meeting Wheel phase 6

Goodbye!

- End the conversation as soon as you have reached a decision on further contact.
- Before the exhibition, prepare different ways to end a conversation.
- Make it a positive send-off so that the visitor will look forward to hearing from you again.

Keep moving in the Meeting Wheel

When the conversation is over, it is time to reach out again. That is your task in the booth: Make visitors feel good and generate leads by having great meetings with the right visitors—over and over again.

Restart

Before every new meeting, you need to do a restart. Remember that even though you, as an exhibitor, will get asked the same questions a hundred times a day, the question will be new to every enquiring visitor.

The world's strongest man, who knew the art of starting over

A few years ago, I had the pleasure of meeting Magnus Samuelsson, who once won the title The World's Strongest Man. It was during a kick-off for exhibitors at an exhibition. He was there to speak and mingle with the exhibitors. During the mingle, it wasn't long before a guy approached him, asking him how he managed to lift that big rock in the world's strongest man competition. Magnus Samuelsson must have been asked that question a thousand times before, but he answered as if it was the first time. "Oh, I'm glad you asked. What's your name—Stefan, let me tell you how it happened. I had to…"

After a while, someone else asked him again and he did the same thing again. It is impressive that the world's strongest man is also the best man at restarts.

Åsa

Practice and improve before every conversation

You are constantly practicing as booth staff. You can't do it perfectly straight away, regardless of how much experience you have…not even if you have read this book from cover to cover. Expect to make mistakes and do not get discouraged. Instead, look at those slips as opportunities to develop and improve.

After every conversation, take a moment to reflect (if you have the time) and give yourself some feedback: "I handled that situation well, but this part I can improve during the next encounter."

What type of person did you just meet? How did you act? How interested were you? What kind of questions did you ask? What was the reaction? What was the body language, the tone of voice and so on? Did it feel good, or was there a nagging feeling that something was missing? If there is a negative in your checklist, what can you do about it?

Remember to help each other improve: give pointers, praise and constructive criticism. And share your experiences, mistakes and successes. Everything will be more fun then.

Give it all you've got

A 100-meter sprinter always aims to run 110 meters. They envision the goal beyond the actual goal, to be able to give it all they have got, even down to the last 2 meters of the race. Give it all you have got. Even during the last, slightly less crowded hour. If the exhibition closes at 6 pm, envision that it closes at 7, so that you will not lose momentum at the end of the day.

The last-hour test

A customer of mine once told me that she had worked as a buyer in the VVS business. She regularly visited exhibitions to find new partners and providers. I asked if she had any tips on how to prepare as a worker in a booth. She smiled and told me that she had found a really great way of knowing which of the exhibitors she most wanted to work with. She formed the habit of visiting the exhibitions during the last hours of the day. She could truly find out who constantly acted professionally. The exhibitors who were still active ended up on her list of approved companies, but the ones who had started packing up, or were browsing their phones, were immediately disqualified.

Åsa

Firstly, you owe it to all the late visitors to be alert. You are hosting your booth and they are your guests.

Secondly, you may end up missing interesting meetings and potential business if you stop reaching out to visitors before the exhibition closes.

Thirdly, it's about your brand. A booth filled with staff sitting down, scrolling on their phones or being generally uninterested, will give visitors a rude and unwelcoming impression, an impression that will stay in every visitor's mind.

After the Exhibition...

Phew! You did it. You all did it. Finally, you will get some hours of sleep, and then deal with all the emails and meetings that have been put on hold because you have been working at the exhibition. Because you are done now, right?

Wrong. The exhibition might be closed but your work is not. Now is the time to harvest.

In this final part, you will find tools and tips on how to use the exhibition as a stepping stone and not as an end point.

Keep your promises

Shattered expectations create a lot of disappointment. Have you experienced disappointment yourself? Perhaps someone has made you a promise and then you find out that you are not getting what you were looking forward to. The person doesn't contact you as agreed. The product isn't up to expectations. Or it didn't appear at the promised time. Not a good feeling, is it? That's the feeling the visitor gets when you make promises you do not keep. So, no matter what: Keep Your Promises!

Schedule the follow-up

Everyone who is involved in following up on contacts must formally set aside time for this important task. Otherwise, you may go back to your everyday work assignments and be swamped by the humdrum tasks. Allotting time for the follow-up is definitely one of the most important factors of success.

You must also constantly check on how the follow-up is going. If you, for some reason, can't keep a promise you made at the exhibition, make sure to give a heads-up and put things right with the help of your colleagues.

Think of the trade show as a qualifying heat in athletics

Be sure to change your attitude towards the trade show. It is not the finish, but the start. The most successful exhibitors can make business and create relationships for the entire year based on leads from the exhibition. A trade show is like a qualifying heat in athletics. If you succeed there, you have made it to the final in 10,000 meters. Now you are truly in the race.

When the exhibition is over,
you have made it to the starting line.

In Conclusion

We want to send you a warm thank-you for reading our thoughts and tips on how to create better meetings at trade shows. Our goal all along has been to make a practical self-help book that you directly benefit from, in your role as booth staff. We hope that you have experienced this and that you feel encouraged and better prepared. Please share the book, and the ideas you like, with your colleagues if they are to be part of the booth staff.

There is a difference between knowledge and competence

Remember that no book in the world can make you more competent, even if it provides you with inspiration and knowledge. It is not until you act and practice that you can build your competence. So here comes one more tip:

Role-play with your colleagues. Act out different situations with different types of visitors. One will be the exhibitor, one the visitor and one the observer. We know it sounds artificial, but it's preferable to meeting a customer without any kind of practice! You are of course welcome to contact Antoni Lacinai to explore the possibilities for a real training program or an inspirational lecture. Many organizations have done so with great success.

In the back of the book there are several checklists that you, and also the project leader for the exhibition participation or the sales/marketing director, may find useful.

The more people you talk to, the more fun you will have at the exhibition. We hereby wish you a really fun exhibition experience with lots of talk, laughter and new contacts!

Good luck!

Appendixes

Appendix 1
Behaviour Bingo

Visit an exhibition. Check a box every time you meet an exhibitor with exactly that good behaviour. Remember to praise them for it!

Which good quality is most common? Which behaviour seems to be the hardest to display? What can you do to "fill your Bingo card" the next time you have a booth at an exhibition?

- Friendly
- Active
- Welcoming while greeting visitors
- Not too pushy or forward
- Treats all visitors well
- Filled with energy
- Listening
- Empathetic
- Curious and interested
- Knowledgeable
- Has arguments that meet my needs
- Articulate

- Trustful
- Focused
- Explains easily so that I understand
- Committed
- Helpful

Appendix 2
What do you need to prepare for?

This current situation analysis will help you find out what you need to focus your preparations on doing a really good job in the booth. Test yourself and your colleagues who are working at the exhibition.

YES | SO-SO | NO

- Do you know why you have a booth at this trade show? What is the purpose?
- Do you know what goals you have during this exhibition?
- Do you know what your task is in the booth?
- Can you explain what is unique about your products and solutions?
- Have you got to know one another? Do you know your colleagues' names?
- Do you know what special skills they possess and when it is suitable for you to guide the visitor to them?
- Do you know how to make a good first impression?
- Do you know what to do to keep yourself filled with energy throughout the entire trade show?

- Do you know exactly which visitors you want to spend some more time with? And who you should just treat nicely? Do you know how to find out who you have in front of you?

- Do you know how to make contact with a visitor without being too pushy?

- Do you know how to get the information you need to be able to review your solution?

- Do you know when it is time to end the conversation and make the interested customer commit to a next step?

- Do you know what to do after the exhibition to keep your promises? Have you set aside time to do this?

Appendix 3
What To Find Out Before The Exhibition—Checklist

Do you have all the information you need to be able to do a good job in the booth? Go over this checklist to make sure that you have not missed anything!

- Do you have new products that you are introducing? What are they? What do you need to know to be able to present them well?
- Do you have a new slogan, a new brand, a new service...or something else whose background you should know?
- Have you invited any VIPs? Who has RSVP'd?
- What is your policy about journalists coming to your booth to interview you?
- What policy do you have concerning competitors that visit your booth?
- What policy do you have for people visiting your booth, who are not in your target group?
- Will any of your top executives be there?

- Is there a plan on how to treat press, analysts and other groups of visitors that are not regular clients?
- Are there other things going on around you that you ought to know about? Lectures, conferences, competitions, and so on?
- What does the trade show look like? Where are other key companies in the field located? Where are the restaurants, cafes and toilets?

Appendix 4
A Checklist For The Project Manager—How To Prepare Your Booth Staff

Here are some of the most important things to think about when you, as the one in charge of the exhibition, prepare your staff for their important task in the booth: to create as many and as good meetings as possible.

At the beginning of the project

- Appoint booth staff in good time—depending on what goals you have for the exhibition; different kinds of people and areas of expertise might be needed.
- Constantly send them updated information about the project.
- Let them join in and share opinions on certain things, for example, "I'm weighing between having this and that giveaway, what do you think?", "What questions do you think we should have in our quiz?", "What

kind of restaurant do you want to eat at on Thursday night?"

- Share information internally (at meetings, via the intranet, in the staff magazine) about your participation in the exhibition and about who will be working in the booth.

The kick-off a few weeks before the trade show

- Practical information
- Information about the exhibition, the placement of the booth and its design
- Travels and lodging
- Lunch coupons (or whatever)
- Booth clothes
- Booth schedule—not more than two hours without a break
- Remind them of comfortable shoes—preferably two pairs
- Remind them to regularly drink water at the exhibition—or they will dehydrate easily

The exhibition's purpose and goal

- Tell them why you are exhibiting (purpose).
- Go over the specific goals you have set.
  Examples: We will get 20 new leads. We will have 200 people testing our product. We will show our new service to 50 of our current customers.

- Agree upon what these goals mean for the booth staff's work at the exhibition.

Example: We will get 20 new leads. To find these leads among all the visitors, you should contact at least seven people every hour.

Training in meetings

- Begin by emphasising that the booth staff are your most important resource at the exhibition. They are the ones who create the meetings which will increase the value of your presence at the exhibition.
- Remind them that their task is to create as many meetings as possible—good ones.
- The booth staff need to have welcoming body language, in order to make the visitors want to stay and make contact. Remind them that they should not talk intensively with each other, that they cannot use their phones in the booth, that they are not allowed to eat in the booth and that they cannot lean on furniture or sit down in the booth (unless they are sitting down, talking to a visitor).
- The booth staff should be active and reach out if the visitors themselves do not make contact.
- Prepare the booth staff to ask open questions: questions that cannot be answered with a Yes or No, to ensure that the visitors stay in the booth.

For example: What company are you from? What do you work with? What interests you here at the exhibition?

- Decide together on questions that the booth staff should ask at the beginning of a conversation, to quickly get an understanding of who the visitor is.

- Review what different kinds of visitors you will meet during the exhibition: current customers, interesting prospects, someone who is not interesting at all as a customer, chatterboxes, and so on. Discuss what strategy you should have for each one. Remember that all visitors are to be treated with interest and respect.

- Practice how you can describe your products and offers in a quick and efficient way. A visitor is often in a hurry to see more of the exhibition.

- Underline the importance of documenting all the visitors who may be interested in having further contact after the exhibition and tell the booth staff how to do this.

Appendix 5
A checklist for planning your exhibition participation

This very short checklist is for you, the project leader in charge of your presence at the exhibition. We could write a whole other book on this subject alone, but here's a quick version:

Purpose and goal

Even if the booth space is already booked, take a step back and determine why you are participating at the exhibition (the purpose) and what you want to accomplish that is tangible and measurable (the goals). NB: It is fine to have several purposes—if you have clearly identified them.

Target groups

Determine what target groups you have:

- Who do you want visiting your booth?
- Who do you want to tell that you are exhibiting?

The common thread

What is the common thread that runs through your participation in the exhibition? What is it that your target group will recognise before, during and after the exhibition? The common thread is:

- Your message
- Your theme (which should be visual, should be suitable for repetition on different media platforms, and should be relevant to your business and your offer.)

Activities in the booth

A good activity attracts the right target group, enhances your message and makes it easier to create meetings and to interact with the booth staff.

Examples of activities are:

- Something to eat or drink
- Trial your product/service
- A competition or challenge
- Mini-seminars in the booth
- Create something
- Hand out giveaways (bags, pens, etc.)

Also, think of a way to help the visitors spread what is going on in the booth on social media!

Invitation

Invite all your target groups, partly because they will feel valuable, but also because you will end up on their "must meet"-list. Remember to:

- Be as personal as possible
- Give them as many reasons as possible to visit your booth
- Try to spread your invitation on social media

Press and media

Have a plan of how you can hand out useful material to the press and media. Take time to find the right angle—something newsworthy. The simple fact that you have a unique or new product is not usually interesting enough—hot enough—for the press.

- Write and send press statements ahead of time, and follow with a phone call
- Arrange a press conference in your booth
- Leave press statements in the press room at the trade show

The booth

The booth should clearly reflect who you are and how you can be of use and it should naturally help you reach your goals. Even though it is tedious, read the exhibition organiser's rules for exhibitors and review what deadlines are set for different orders.

Here are some questions that are good to ask oneself at an early stage:

- that kind of meetings will take place in the booth? How do we want the visitors to move? What should the flow look like?
- What space do we need for activities?
- Do we need to hang something in the ceiling to be more noticeable?
- Do we need to order more power outlets than the ones we have?
- What kind of lighting do we need?
- Do we need to install water in the booth?
- Do we need space for storage?
- What help do we need from the outside regarding the design, the construction and the lighting?

Booth staff

Make sure that your booth staff have the right competence and are properly prepared for the exhibition. (Checklist in appendix 4).

Following up on the contacts

It is just as important to follow up on the contacts after the trade show. Even if you, the project manager, can relax a little after the exhibition, it is vital that you keep up the pace until the last contact has been followed up on.

- Appoint someone to be responsible for the follow-up
- Set a clear deadline for when each lead should be contacted
- Reward those who do a good job with the subsequent work

Evaluate the exhibition participation

- What results did you get? Did you fulfil your goals? What other, unexpected, results did you get?
- What led to the results? Was it your own doing? Was the organisation of the exhibition better/worse than you expected? Were there other factors?
- What can you learn and take with you to the next exhibition? Write a brief report.

Appendix 6
The 10 Commandments
for booth staff

1. You shall always reach out to a visitor, because they are there to see you.
2. You shall never talk to your colleagues so intensely that the visitor feels left out.
3. You shall put away your phone and only use it when you are not in the booth.
4. You shall not have your lunch in the booth.
5. You shall listen to your visitor as you would like to be listened to.
6. You shall sleep properly the night before the trade show.
7. You shall not sit in the booth.
8. You shall honour and welcome all your visitors—no one should feel dumped just because they are not your dream prospect.
9. You shall always wear comfortable shoes.
10. You shall drink water and take breaks so that you can endure until the end.

"Unlock the art of magnetic booth encounters at trade shows: Master the dance of engagement, from the spark of the first impression to the lasting glow of a promise kept. This guide is your blueprint to transforming every interaction into a stepping stone for success, ensuring you're not just participating but captivating, not just present but unforgettable. Dive in to discover how every handshake can turn into a milestone for your business."

Ksenija Polla – Head of education and legacy programmes. ICCA[1]

---

[1] FYI: (ICCA stands for the International Congress and Convention Association)